Belgian, Dutch and German Chicken Breeds
A Guide to Belgian, Dutch and German Fowls

by International Correspondence Schools

with an introduction by Jackson Chambers

This work contains material that was originally published in 1912.

This publication is within the Public Domain.

This edition is reprinted for educational purposes
and in accordance with all applicable Federal Laws.

Introduction Copyright 2017 by Jackson Chambers

Self Reliance Books

Get more historic titles on animal and stock breeding, gardening and old fashioned skills by visiting us at:

http://selfreliancebooks.blogspot.com/

Introduction

I am pleased to present yet another title in the "Chicken Breeds" series.

This volume is entitled "Belgian, Dutch and German Chicken Breeds". It was originally published in 1912 by the International Correspondence Schools as "Belgian, Dutch and Germn Fowls" and was part of a correspondence course offered on Standard Breeds of Poultry.

Included in the course are details on such well known breeds as the Campine and the Hamburg, but also lesser known breeds such as the Herve and the Malines Fowl.

The work is in the Public Domain and is re-printed here in accordance with Federal Laws.

Though this work is a century old it contains much information on poultry that is still pertinent today.

As with all reprinted books of this age that are intended to perfectly reproduce the original edition, considerable pains and effort had to be undertaken to correct fading and sometimes outright damage to existing proofs of this title. At times, this task is quite monumental, requiring an almost total "rebuilding" of some pages from digital proofs of multiple copies. Despite this, imperfections still sometimes exist in the final proof and may detract from the visual appearance of the text.

I hope you enjoy reading this book as much as I enjoyed making it available to readers again.

Jackson Chambers

BELGIAN, DUTCH, AND GERMAN FOWLS

Serial 1376   Edition 1

BELGIAN

INTRODUCTION

1. The fowls most popular in Belgium are the Braekel, the Campine, and the Malines. Other varieties that attract more or less attention in that locality are the Bruges, the Brabant, and the Herve. All of these except the two first mentioned are shown in Fig. 1. This illustration shows, also, the Drente and the Owl-Bearded Dutch, which are Dutch varieties; the La Bresse, which is a French variety; and the Lakenfelder, which is a German variety. During the past few years the Belgian, Dutch, and German fowls have come into increased notice. A German writer claims that Belgium has twelve distinct breeds; Holland, nine; and Germany, twenty-three. Although but few fowls of these breeds have been bred in America or England, they are gradually coming into favor.

The most popular of all of the fowls mentioned, outside of their own countries, are the Campine and the Lakenfelder. The others have received very little attention in America and but little more in England, the Lakenfelder excepted. Of all the fowls included in the group, the Malines and the La Bresse have been most developed. Following these, the Lakenfelder and the Owl-Bearded Dutch have been devel-

Fig. 1

oped in the order mentioned. Recently, the Belgian breeds have been brought into public notice, and as a result many of them have been taken into France, England, and the United States, where they are being successfully bred.

BRAEKEL

2. Origin.—It is thought that fowls of the same general characters as the **Braekel** existed in many parts of Europe as early as the 12th century. It can scarcely be doubted that they have descended from the same type of fowl that produced the Penciled Hamburg; and that they are largely Italian is indicated by their general breed characters and the style of their comb and ear lobes.

3. Development.—The Braekel fowls have been developed in that part of Belgium where the conditions are most favorable to poultry growing. Careful breeding has developed them into larger fowls than the Campines, and for this reason they are better for general purposes.

4. History.—The Braekel fowls are bred in the western part of Belgium and in some parts of France. They have been carefully bred for both egg production and market poultry, but not so carefully bred for exhibition. Braekel-Campine fowls were brought from Belgium to America and were admitted as two varieties to the Standard of 1894; they were dropped from the Standard of 1898. They were called Silver Campine and Golden Campine, the back of the male of the silver variety being a silvery white, and the back of the male of the golden variety being darker than golden bay.

5. Description.—In body formation, the Braekel is somewhat like a Leghorn or a Spangled Hamburg of large size. The official Standard of Belgium says that the comb, face, and wattles of both male and female shall be red; beak, blue or horn colored; ear lobes, almond shaped, mother-of-pearl color for the male and either white, bluish white, or mother-of-pearl in the female; head of fairly good size; comb of male, large, single, with five or more points, that of the female falling over to the side like the Leghorn comb; breast, deep and full; body, broad and of medium size; back inclined slightly toward the tail; tail of the male well spread, sickle feathers about one-half longer than the main tail feathers.

In color, the hackle, wing bows, and saddle of the male of the silver variety are white, with some specks of black scattered through them; the rest of the body plumage should be barred, the barring extending into the sickles and coverts of the tail. The hackle of the female is white, and the rest of the plumage is barred, the barring extending into the tail feathers. The main tail feathers of both are more or less spotted with white. The distinguishing features between the Braekels and the Campines are the white back of the Braekel male, the light-colored breast, and the very dark main tail feathers of the female; both the male and the female Campine is barred over the entire body, with no white on the back of the male.

Silver Braekel fowls are mixed white and black. The black bars of the body plumage should be at least three times as broad as the white; the black should glisten with sheen, and the white should be of a grayish tint rather than pure white. The golden Braekel fowls are of a golden bay and black color. Other varieties of Braekels have been bred in Belgium, the most prominent of which are the Chamois and the Blue. The color of flesh and skin in all varieties is white; the shanks and feet are slaty blue. Males average from 6 to 8 pounds; females from 4 to 6 pounds, according to age.

6. Mating.—In Belgium, in selecting Braekel fowls for the production of offspring of proper body formation, care is exercised to mate only such fowls as have a plump body and a long, full breast that carries a large amount of breast meat in proportion to the size of the fowl. When mating for variety colors, males and females that conform in plumage, shank, and toe coloring are selected; white top coloring in males of the silver variety is considered highly desirable. Otherwise, Braekel fowls selected for mating are much like the Campine fowls.

BRUGES

7. Origin.—The **Bruges** fowls have been bred in the northern part of Belgium, in a province of the same name, for many years. No one seems to know just how they were originated. Claims are made, however, that they were produced from a cross between the Malay fowls brought from India and the fowls common to Northern Belgium.

8. Development.—The Bruges fowls have been but little cared for until recently, when they came into slight notice in other localities, and the poultry growers of Belgium have developed them more for sale as new breeds and varieties rather than for any other purpose.

9. History.—No definite knowledge as to the history or early existence of the Bruges fowls can be gleaned. Up to recently they were but little known outside of their native district in the northern part of Belgium.

10. Description.—There are four known varieties of Bruges fowls: red, black, white, and blue. All have the general appearance of the Malay family, yet they have the same traits that dominate the poultry of Belgium: the white flesh and skin, the leaden color of shanks, and the body formation general among market poultry. The comb, wattles, and ear lobes of these fowls are red; the combs are rather small. In weight, the males average from 8 to 10 pounds; the females, from 7 to 9 pounds, according to their age. They are now bred more like the Cornish Game fowl than formerly.

11. Mating.—There is no special rule for the mating of Bruges fowls other than to select the best in size, shape, and color, with uniform color of shanks, and then to breed their offspring for improvement.

CAMPINE

12. Origin.—The Campine fowls have descended from fowls which were mentioned by Aldrovandi as Turkish fowls. It is impossible to state definitely the place of their origin, but it is probable that the Campine, the Braekel, and the Penciled Hamburg fowls have descended from Italian fowls that were scattered throughout the populated districts of Belgium and other near-by countries.

13. Development.—In Belgium, the Campine fowls have been developed principally for egg production. Some Campines were brought into England about 1885, and since then they have been developed in England for exhibition purposes. Two varieties, the Golden and the Silver Campine, have been thus developed. Both of these varieties are bred in America.

14. History.—In Belgium, the name Campine is given to the smaller fowls that are found in the districts of La Campine. The larger fowls of the same type found in other parts of Belgium are called Braekel. The source of both the Campine and the Braekel is identical, but generations of breeding from different lines and under different conditions of nutrition have caused them to separate. Experts will at once distinguish between them; those not so well informed notice only a difference in size. A Campine Club was formed in England in 1899, and through its influence a marked improvement in exhibition qualities has been made. Modern Campine fowls are bred in America, and a club was formed in their interest during the spring of 1911.

15. Description.—Campines are single-comb fowls with the general appearance of the Penciled Hamburgs. They are larger than this variety of Hamburgs, and both varieties have a darker shade of body color. The English Club Standards requires that the plumage of the male, including the tail, be barred or marked the same as the female of the variety to which it belongs. The ideal Campine fowl of the silver variety has a silvery-white neck hackle and barred black-and-white body plumage. A fowl of the golden variety has a golden hackle and barred black-and-golden body plumage. The golden color of fowls of this variety is a yellowish bay, in contrast with the reddish bay in the color of the female Golden Penciled Hamburg. The black bars should be three times as wide as the white bars in the silver variety and as the golden bars in the golden variety. The tips of the feathers should be white in one variety and golden in the other. The bars should be transverse and distinct, and the lines marking the divisions between the bars should be straight and regular. Straightness and regularity of the bars is of more importance than their direction. The black should be rich in color and have a green sheen, and each color should be pure. The eyes should be dark; the comb, face, and wattles, red; the ear lobes, white. The comb of the male must be upright and of medium size, neat, and well serrated; the comb of the female should be of medium size and should fall over, or be inclined to fall over, like the comb of the Leghorn. The shanks and toes of both varieties should be leaden blue; the beak, horn colored; and the toe nails, dark or horn colored.

The Silver Campine has a white ground color and black bars; the Golden Campine has a golden ground color and black bars. The difference in these two varieties is the ground color, which is white in one and golden bay, much like the body color of the Golden Penciled Hamburg females, in the other.

The chief beauty of the Campines is their attractive form and clear color. A male of either variety that is lacking in richness of sheen on the black bars is of but little value. The hackle of both the male and the female of the silver variety must be pure white and free from spots of black, brown, or any foreign color. The main tail feathers of both male and female should be more or less barred. The sickles of the male should be long and have an attractive sweep, or curve, and both the sickles and the coverts should be barred with black. Gray spots or faint bars across the black bars must not appear in either the male or the female of the silver variety. The white must not run into the black bars of the Silver Campine, nor must the golden-bay color run into the black bars of the Golden Campine.

16. Mating.—In selecting Campine fowls for breeders, both the male and the female should possess the best Campine type that can be selected, and special attention should be given to selecting females of medium size and males of fairly large size for the breed. The females selected should possess good form and be prolific layers of large eggs. The plumage color of both should be pure and well defined, the black having a rich green sheen, and the tip of each feather should be well marked. The beauty of these fowls depends largely on the bars appearing like rings about the body. The regularity of these rings is broken by the shape of the tip of the feather. Fowls having horseshoe markings on the breast should not be used for breeding purposes.

In selecting breeding fowls for mating for the production of Silver Campines, those with a white ground color should be selected; the black bars should be almost three times as wide as the white bars. The neck hackle of both males and females should be white. The rest of the body of the male and the female should be barred. There is a tendency for males to have white backs and saddles. To overcome this, only males of the proper color and barring and that have dark under plumage should be bred from. To intensify and make more brilliant the dark bars, only males and females having dark color in the under plumage in the back, breast, and body should be bred from.

In mating the Golden Campine, the body plumage color of the breeding fowls should be a rich golden bay throughout, barred with black; the neck hackle, golden bay, free from barring and striping, and pale or yellow color. Otherwise they should have the same general breed characters and color markings as the Silver Campines. The under plumage of the Golden Campine should be darker than the surface plumage. Golden Campine fowls have dark under plumage, but to secure the best surface color, golden bay next to the skin, darkening into almost black at the upper edge of the fluff, is desirable.

HERVE

17. Origin and Development.—The origin of the **Herve** fowls is hidden in obscurity. The only indication of their ancestry is their Hamburg-like appearance. These fowls have been so little bred that it cannot be said that they have been developed beyond a meager extent. Fowls of this type have been bred in Belgium for a number of years. They are but little known outside of their own territory, and are mostly sought after by amateur exhibitors who are pleased with their small size and attractive plumage colors. There are three varieties: the Black, the Blue, and the Cuckoo Herve. The Blue Herve is the most beautiful variety.

18. Description.—The Herve is a fowl of small size not much larger than a bantam, and has the general appearance of the Hamburg. The Blue Herve is marked like the Andalusian; the Black Herve is of brilliant color and rich in sheen; the Cuckoo Herve is marked like the Malines; the Black Herve has black shanks and feet; the Blue Herve has lead-colored shanks and feet; and the Cuckoo Herve has pinkish-white color in the shanks and feet. The males weigh from 3 to 4 pounds, the females from 2 to 3 pounds. They have single combs. The color of the combs, wattles, and ear lobes is red. The males have beautiful, long, flowing sickles.

19. Mating.—In mating these fowls, the same care in selection must be given to the Black Herve that is mentioned for the Black Hamburg; the Blue Herve fowls must be mated the same as are Andalusians; and the Cuckoo Herve fowls the same as the barred varieties. In all varieties, the shanks and toes of the males and the females must match in color.

MALINES

20. The **Malines** fowls of Belgium are bred for market. They are large and have gained a wide reputation under the name of Poulardes de Bruxelles. Many varieties of Malines are recognized in their own country, and in England one variety has been accepted as an exhibition fowl. The varieties most plentifully bred in Belgium are the Black, the Ermine, the Cuckoo, and the Turkey-Headed Malines. The Turkey-Headed Malines is so called because of its peculiarly shaped head with a triple comb, such as is shown in Fig. 2. The Cuckoo Malines is the variety that has been most commonly bred for exhibition.

Fig. 2

21. Origin and Development.—A fowl that resembles the Malines has been known for several centuries. In recent years, some fowls of the original Malines type were crossed with Langshan and Antwerp Brahma fowls, and offspring were selected of a color resembling the Barred Plymouth Rock. The Turkey-Headed Malines came naturally from the Langshan-Brahma crosses.

The Malines fowls have been developed into large size and plump body formation, primarily for competing with fowls of other countries in the markets of France and England. Although the fowls of the various varieties of this breed have done remarkably well in their native country, they have not done so well in other countries, and in America have not proved to be any more satisfactory than the native fowls.

22. Description.—The head and neck of the Malines fowls, though strong and well proportioned, are not excessively large. They have the least amount of waste of any fowls of their size. In breast and body formation, they are broad, deep, and full; the back is long, flat, and broad across the loins; the thighs are strong, well proportioned, and set well apart, and the breast hangs low between them. The body formation of the Malines is said to be almost square, and, when finished for market, the fowls have approximately that shape. In carriage and general appearance, the Malines are not unlike the Brahmas. According to their sex, they weigh from 8 to 10 pounds at the age of 10 months.

The color of the Cuckoo Malines for exhibition is: Beak, white or horn colored; comb, face, wattles, ear lobes, and eyes, red; shanks and toes, white. The plumage resembles that of an indifferently marked Barred Plymouth Rock.

23. Mating.—In mating Malines fowls for the production of exhibition offspring, general breed characters should have the greatest consideration. The rules for color and mating for Barred Plymouth Rocks should be applied to fowls of the Cuckoo variety, and those having the desired color should be separated and mated. Where body proportions only are desired, fowls that are not lacking in breast and body formation should be selected and mated.

In selecting to produce black offspring, Malines of the proper form and size and having perfectly black plumage must be selected. The Ermine, or Light Malines, which has plumage color like the Light Brahma, should be selected for Malines shape and color resembling the Light Brahma. Fowls of pure white plumage must be selected to produce the White Malines, which is a new variety.

MISCELLANEOUS BELGIAN

24. Other breeds of Belgian fowls are the *Antwerp Brahma*, the *Ardenne*, the *Brabant*, the *Flemish*, and the *Huttegem*. The fowls of these breeds have been bred in comparatively small numbers.

25. The **Antwerp Brahma** has been bred in Belgium mostly for crossing with other fowls, and has never been bred for exhibition. It has been described among the Asiatic fowls.

26. The **Ardenne** fowls are formed somewhat like the game fowls. They have single combs of medium size, and their plumage is somewhat like that of the Black-Breasted Red

Game fowls. The females are darker in plumage than those of the game varieties. In weight, they average from 1 to 6 pounds, according to their age and sex.

27. The **Brabant** fowls are not unlike Houdan fowls of inferior quality. They are large and of heavy body formation. There are several varieties, the principal ones being the Black, the Speckled Black, and the White. Coming as they do from the locality where Hamburg, Crevecœur, and La Flèche fowls have been bred, it is not to be wondered at that a fowl having the crest, beard, and comb of the Crevecœur, the body formation of the Braekel, and the color and markings of all should be produced.

28. The **Flemish** fowls are cuckoo colored, or barred, like the Barred Plymouth Rocks. Were it not for the peculiarities of white flesh and skin, and pinkish-white shanks and toes, they might be called the Barred Plymouth Rocks of Belgium.

29. The **Huttegem** fowls are thought to have descended from offspring created in developing the market poultry of Belgium. They are bred in three variety colors, the Gold, the Barred, and the Light Brahma. They have white skin and flesh, and bluish-white legs. They lay tinted-shelled eggs, which indicates the Asiatic blood in their make-up. They are somewhat like the Malines, and have both single and rose combs.

DUTCH

HAMBURG

30. Origin.—The **Hamburg** fowls are the most important Dutch fowls. Aldrovandi, an Italian naturalist, was the first to mention them. His first writings relative to this breed were published in 1599, in which he called them *Turkish*, and his description of them, though not complete, stated that they were white, spangled with black spots, and that their feet were tinged with blue. Another hen was described as one of the same color, except that instead of being white, she was of a yellowish color, spangled with black, her feet being blue. Although this description is rather incomplete, the illustrations, consisting of wood cuts, that accompanied it, indicated that these fowls were the forerunners of the present Hamburg fowls. Two hens, one golden and one silver, shown by these old illustrations, were clearly marked with the peculiar fleshy rose comb of the Hamburg, terminating in a sharp point behind. Mr. Dixon states that this comb is seen in no other variety of fowl, and that it is well described by *apicem in vertice gerit*, which translated into English means, "It carries a wisp of flame for its crest."

Edward Brown, of England, states that Hamburg fowls as we now have them are traceable to two distinct sources. The two Spangled Hamburg varieties and the Black Hamburg variety originated in Great Britain and have been bred in the North of England for two centuries, if not longer; the Penciled Hamburg came from the Netherlands. These fowls have been changed considerably since their introduction. It is necessary to deal separately with the Spangled, the Black, and the Penciled Hamburg varieties. Mr. Brown further states that the peculiar comb of the Hamburg is not found on fowls of any other race that can be traced, and that the assumption is not unreasonable that the British Spangled Hamburgs are descendants of the fowls described by Aldrovandi.

The claim has been made by W. B. Dickson, of England, that the Hamburg is a variety of the Paduan (Polish) fowl. The same opinion appears in the early writings of Bonington Moubray, who wrote of poultry prior to the 18th century. In describing the Polanders (now the Polish fowls), he wrote: "They are sometimes called Everlasting Layers." This was one of the early names for the Hamburg fowls. Undoubtedly, the modern type of the Black and the Spangled Hamburg was made in England, and that of the Penciled Hamburg varieties in Holland. Evidently, all Hamburg fowls originated in Italy. Those that came to Britain in the early days were freely bred in the northern counties of England, and the Black and Spangled Hamburg varieties were derived from them. Others were taken to Holland, and we find recorded by Bonington Moubray, prior to 1816, the following statement: "Besides the Polanders, there is a small variety now imported from Holland called 'Every-Day Layers,' which are everlasting layers." From these the Penciled Hamburg varieties have been developed.

As previously mentioned, the Spangled Hamburg was developed in England. In writing of them, the Rev. E. S. Dixon, A. M., of Norfolk, England, says: "This beautiful variety is distinguished from other members of the same family by its large topknot being colored instead of white, and by the black and conspicuous muffle, or ruff, on the throat and under the beak. There are two kinds of them, the golden spangled and the silver spangled, the ground of the feathers of the former being a rich yellow, approaching an orange red, with black spots or spangles; the silver spangled differs from the preceding by having the ground of the feathers a silvery white." The early illustration of the Hamburg fowls described by Dixon indicates that they might have belonged to the Paduan rather than to the Turkish breed mentioned by Aldrovandi. In writing further of the Hamburg fowls, Mr. Dixon states that in the neighborhood of Keighley, which is in Yorkshire on the border of Lancashire, the Bolton Grays are called Chittaprats. He also states that they were known as Bolton Grays, Golden Pheasants, and Every-Day Layers.

BLACK HAMBURG

31. Origin.—The **Black Hamburg** fowls were bred and exhibited as Black Pheasants in about the year 1800, when they were admitted at a village show in Lancashire. At that time they were thought to be of pure Hamburg origin. It is believed that they originated from the black offspring of full-tailed Silver Moonies. They were crossed with Black Spanish to improve the size and white in their ear lobes. The cross was detrimental, from the fact that white appeared in the face of many offspring, and their heads were coarse.

32. Development.—The modern Black Hamburg has been developed from the original Black Mooney. The quality of these fowls has resulted from long continued selecting, mating, and breeding for perfection of quality and beauty of plumage. The Black and the Spangled Hamburg varieties are so nearly related as to make their history and development almost identical.

33. History.—In writing of the Hamburg fowls, Charles Holt, of England, Honorary Secretary of the Hamburg Club, states: "They were exhibited long before the Birmingham show was known; they were then called Black Pheasants, and were exhibited for what I think must now be called the celebrated 'copper kettle.'" This is supposed to have been a hanging kettle that was used at that time as a champion cup for all Hamburg fowls. The meetings were held in the taverns of the town. The birds were brought in cages much like those used now for parrots. The exhibits were usually held during a half holiday and continued into the evening. All exhibitors were judges; they decided among themselves which were the best, and the awards were made with good feeling. A feast and jollification usually followed the decision. A picture of an early Hamburg show in which the copper kettle may be

Fig. 3

seen is shown in Fig. 3. Later, the Hamburg fowls were taken up by fanciers.

34. Description.—An idea of the general contour of all Hamburg fowls can best be gained by reference to the color plates. Comb and head points are of even greater importance in the Black Hamburg than in other varieties. The beautifully formed comb and head points, with their rich, brilliant color, are like a crown of scarlet and white adorning a fowl of beautiful formation and of a plumage color of glistening black throughout. A beetle-green sheen overcasting the entire plumage brings out the beauty and richness of the black, and in contrast with the beautiful sheen is the scarlet color of comb and the head points embellished with large, snowy-white ear lobes as soft as kid. The ear lobe must be enamel white, soft, delicate, and well placed against the side of the face. The shanks and toes must be leaden blue; the eyes, red; and the beak, black or horn colored.

35. Mating.—In mating Black Hamburg fowls, the best success will be obtained by breeding in line for many years for the purpose of establishing form and color. To do this, the finest fowls that can be found must be mated, and their offspring remated. This process must be followed continually without the least deviation. Henry Pickles, of Earby, England, who was, perhaps, the best breeder of Hamburg fowls of his day, stated that he had maintained a mating of them for 25 years. He said: "When the cock failed me, his best son took his place. When one of the hens failed me, her own best daughter took her place; and in this way for more than 25 years I have bred them better and better each year, until the extreme has been reached, and no more color can be permitted." In mating for black, males and females of the richest color with the greatest amount of sheen should be mated until red makes its appearance in the plumage of the offspring. When this occurs, hens of somber-colored plumage must be introduced to overcome the influence of the red.

36. Diagram for the Mating of Hamburg Fowls. The breeding of no one breed or variety has had attention equal to that given to the breeding of Hamburg fowls in England. In writing of successful methods, the Honorary Secretary of the Hamburg Club of England states that the principle shown in the diagram of Fig. 4 is the best that can be adopted for the mating of these fowls, as the method can be followed year after year with continued success, and it might well be used for mating fowls of other breeds.

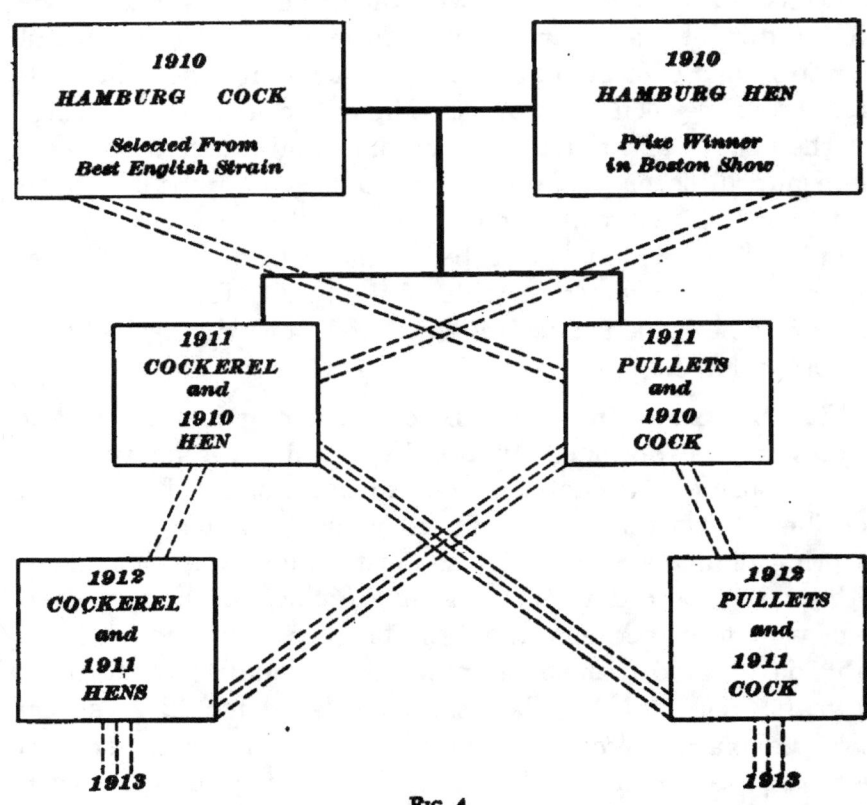

Fig. 4

The diagram is somewhat like the mating chart shown in *Standard-Bred Poultry*, Part 2, the main difference being that the fowls mated the first year are connected by a single black line; those mated the second year are connected by two dotted lines; and those mated the third year are connected by three dotted lines. This process could be continued for as many years or generations as might be necessary. It will be noticed that the foundation of the strain was a pair of carefully selected fowls.

It is thought that 4 years of breeding in this way will establish a strain of fowls that will have superior breeding qualities. This will be true, of course, only when the original fowls are perfect and the female is a producer as well.

In order to be successful, a fancier must not only be familiar with the method but must be expert in selecting each year for mating the best fowls for the purpose intended. It is necessary, under this system, to keep accurate records and to mark the fowls by toe marks or identification bands.

GOLDEN PENCILED HAMBURG

37. Origin.—Prior to 1785, penciled fowls came to England from Holland, and thence to America. Before being divided into separate classes, they were known by many confusing names. Moubray called them Every-Day Layers. When they first came from Holland, they were called Penciled Dutch; as they were scattered throughout the world, they were known as Bolton Grays, Bolton Bays, Creoles, Creels, Chittaprats, and Pheasants. It was not until the fowls had been separated into classes that the Hamburgs became known as a distinct breed and a standard was established for them. There were at first two varieties of the Hamburg, the light and the dark, which were later known as the silver and the golden. Fowls of the golden variety were selected and bred until they had been produced with color and markings so distinct as to leave them without rivals, and they were then named **Golden Penciled Hamburg.**

38. Development and History.—The Golden Penciled Hamburg fowls have been developed by establishing strains as described for the Black Hamburg. The chief aim has been to separate the golden bay from the silver, and to have the barring of the female so regular as to give it the appearance of having been laid on with geometrical precision, and the males ideal in color and free from foreign color.

The history of all Hamburgs as exhibition fowls can be told best in connection with the Silver Penciled Hamburg.

39. Description.—The general formation of the Golden Penciled Hamburg is shown in the color plate of the penciled variety. In the Golden Penciled Hamburg male, the color of the neck hackle and breast is a bright bay or reddish bay, and the body is reddish bay. In females, the body color is clear reddish bay, each feather being barred with black. The English Club Standards describes them in both sexes thus: Beak, horn colored; eyes, red; comb, face, and wattles, red; ear lobes, white; legs, leaden blue. In the males, the neck hackle, back, saddle, shoulder, wing bows, breast, and upper parts, are a bright-red bay; also the wing coverts, and the bottom web, or visible part of the feather, is generally black or coarsely penciled; the tail is black, tinged with green; the sickle feathers and tail coverts are of a solid, rich, transparent green surface color and black foundation, and are laced all around with a narrow strip of gold. In the female of the Golden Penciled Hamburg the neck hackle is of a bright golden color; all the rest of the plumage of the fowl is of a bright, golden color, each feather being penciled distinctly and evenly across with fine parallel lines of a rich green hue. The penciling and the intervening lines should be of the same width; the secondaries should be penciled as much as possible, but the markings are naturally a little coarse.

40. Mating.—A male that has the finest exhibition qualities, especially good head points, mated with hens of equal quality and from the same strain from which the males have come, will prove to be the most satisfactory breeding fowls for the production of the exhibition Golden Penciled Hamburg. It is almost a waste of time to attempt to breed Hamburg fowls of sterling quality without having stock fowls that are not only excellent in themselves but near akin and bred in line from the best. In breeding for color, the richest-colored fowls, as described in the English Standard, should be selected. The ear lobes on some of the finest males are frequently almost as large as a silver half dollar, as smooth as glass, and as white as the most beautiful enamel. The lobes of the hen should be as large, comparatively, as those of the male. The exquisite

§ 7   BELGIAN, DUTCH, AND GERMAN FOWLS

shape of the Hamburg comb cannot be lost sight of in selecting fowls for breeding. They should have eyes of a rich, bright red; the shanks and toes should be of a leaden color and as smooth as polished brass. More definite information relative to the necessities of mating Hamburg fowls is given in treating of the Silver Penciled Hamburg.

GOLDEN SPANGLED HAMBURG

41. Origin.—The **Golden Spangled Hamburg** fowls descended from about the same source as the Silver Spangled Hamburg. To avoid unnecessary repetition, and because the history of the Hamburg is more closely connected with that of the Silver Hamburg than with that of the Golden Hamburg, a more complete statement of their origin will be found in the discussion of the Silver Spangled Hamburg.

42. Development and History.—Golden Spangled Hamburg fowls have been developed from the early Mooney and Yorkshire Pheasants. Lewis Wright has stated that there was in Lancashire a variety called Golden Mooney. These fowls were smaller than the Silver Moonies, but in color and markings they excelled the latter. It is thought that the Golden Spangled Hamburg fowls have been developed from them.

The history of the Golden Spangled and the Silver Spangled Hamburg is nearly identical and is told in connection with the Silver Spangled Hamburg.

43. Description.—The Golden Spangled Hamburg fowls should be described to meet the requirements of both the American and English Standards. Being a fancier's fowl and having been made and perfected by the fanciers of England, more than passing attention must be given to the difference of description, so as to give full information concerning their dual existence. The same description as to shape will serve for the Spangled, Black, and Penciled Hamburg varieties. Notwithstanding this, they differ materially in their general make-up, the Spangled Hamburg being heavier in body than the Penciled Hamburg fowls.

The Penciled Hamburg fowls are strictly Italian or Mediterranean in form, and more like the Leghorn in their general make-up than like any other fowl. The Spangled Hamburg fowls are larger than those of the penciled varieties and more like general-purpose fowls. The females of the penciled varieties are small in comparison to the modern type of the spangled variety.

The American Standard describes the Hamburgs as of medium size with rose combs of beautiful formation, well serrated, and not so large as to extend over the width of the head on either side. The comb should be square in front, the spike of the comb tapering and extending out behind in fair proportion to the rest of the comb. The back should be of medium length, broad in front, and gradually sloping to the tail; the breast should be broad, the body round, the fluff short, the wings large and carried rather low. The English Standard says that wings are large and neatly tucked up. The American Standard demands that the flowing tail of the Hamburg shall be carried at an angle of 40 degrees and the English Standard calls for an angle of 45 degrees. In both Standards, the description of the hen conforms to the description of the male.

The English Standard describes the comb as square in front, gradually tapering toward the back, and ending with a long spike pointing in a straight line with the surface of the comb. The comb should be firmly and evenly set on the head. The top level is covered with points, and the important feature in this is that the spike shall point in a straight line from the surface of the comb. In the description of head points, the English Standard is less explicit, requiring the head to be neat in formation; the beak is to be short and small; the eyes, full and round; the wattles, thin, well rounded, and free from wrinkles; the neck, of medium length and nicely arched; the hackle, very full, of a good length, and coming well over the shoulders. As to body formation, the breast is described as prominent and round; the back, of medium length; the wings, large and neatly tucked up; the tail, of good length and carried at an angle of about 45 degrees; the sickles and secondaries, broad, plentiful, and sweeping; the thighs, somewhat short but neat; the shanks, small boned and medium in length; the toes, slender and well spread. The Standard requires that the head shall be carried erect, the chest well out and forwards, and that the whole appearance shall be lively and graceful. In the penciled variety, the male should weigh about 5 pounds; in the other varieties, the male should be somewhat heavier. The plumage should be very profuse. Penciled Hamburg females should weigh about 4 pounds; the females of other varieties should be heavier. In other particulars, the hen should conform in an effeminate way to the male.

The color description of both sexes of the Golden Spangled Hamburg, according to the English Standard, is as follows: Beak, horn colored; eyes, comb, face, and wattles, red; ear lobes, white; legs, leaden blue. In the male, the hackle is of a rich golden bay, each feather marked down the center with a stripe of beetle green; the back and saddle, golden bay, almost maroon, with a dagger-shaped tip at the end of each feather. The wing bars should be two in number, consisting of rows of large green spangles running parallel across each wing with a gentle curve, each bar distinct and perfect; the secondaries are golden bay, tipped with large, round, green spangles that form what are called *steppings;* the breast and under-body plumage are of a rich golden bay, and each feather is tipped at the end with a round greenish-black spot or spangle (the greener the better), small near the throat and increasing in size toward the thighs, but never so large as to overlap; the main tail feathers, the sickle feathers, and the coverts are of a rich, transparent green on a black foundation.

In the hen, the head is black and bay mixed; the neck hackle, a rich bay, each feather marked down the center with a green stripe; in the saddle, shoulders, wing bows, breast, and under body, every feather is of a rich bay tipped with lustrous green spangles; each feather should be as long as possible, but never so large as to overlap; the wing bars should be two in number, sometimes three, consisting of rows of large beetle-green spangles running parallel across the wing in a gentle curve, each bar distinct and separate; the secondaries of the wing should be, as in the male; the main tail feathers should be black, tinged with green; the coverts should be spangled. In some instances only a slight lacing of gold is seen around the feathers. With reference to the English Standard, one of the prominent writers of England states that in the Golden Spangled Hamburg, the tail of the hen, and the main tail, sickles, and coverts of the male are a rich glossy green-black without spangles; the hackles of both sexes, and the saddle and back of the males are tipped instead of having spangles. The American Standard has the same requirements for the male, but the coverts of the female are required to have greenish-black spangles.

44. Mating.—In mating Golden Spangled Hamburg fowls, success will depend on the quality of the fowls used in the matings. It will be useless to attempt to produce offspring of even medium quality unless the stock is of the best and selected from a strain the fowls of which produce well and the males and females of which are of the same line of breeding. In the production of pullets, hens of the highest exhibition quality should be mated with the best males that can be obtained from a pullet-breeding strain. From these, a line of pullet-breeding males and females can be bred, and they must be kept separate as carefully as would be required for the production of any of the parti-colored varieties. Cockerels will be produced in much the same manner, except that an exhibition male of the highest quality must be mated with females descended from such a line of breeding. In other words, to succeed well in the production of Golden Spangled Hamburg fowls, they must be line bred from foundation stock of the highest quality, and the matings must be kept separate; new blood must be introduced only through a female of like quality introduced into the male line of breeding.

SILVER PENCILED HAMBURG

45. Origin.—In writing of the origin of the **Silver Penciled Hamburg**, W. B. Tegetmeier, F. Z. S., of England, states that the penciled and spangled fowls, although frequently described together, are so essentially different that they should be regarded as distinct varieties. The white body, the black

§7 BELGIAN, DUTCH, AND GERMAN FOWLS

markings, the greenish-black tail, and the blue-tinged legs are all characters that prove Aldrovandi's knowledge of them when he called them Gallina Turcica, or Turkish fowl. He states also that the silver penciled variety descended from the Bolton Grays, and the golden variety from the Bolton Bays, and that the many names applied to them were of local origin and referred to the original Penciled Dutch, as they were called when they first came to England.

46. Development and History.—The Silver Penciled Hamburg has been developed from the crude originals into beautiful type. The color description of the present in contrast with the color description of early days tells of their development.

The history of the Silver Penciled Hamburg is so closely connected with that of the other varieties as to make it possible to give the history of all under that of the spangled variety.

47. Description.—The description of the originals of the Penciled Hamburg might be compared with the description of the modern variety. The early Penciled Hamburgs were fowls of small size, compactly built, and very active. The body plumage of the females of one variety was white, penciled with transverse bars of black, and the body plumage of females of the other variety was golden, penciled with black. The neck hackle of the silver variety was perfectly free from dark markings; the males were free from these markings, their plumage color being either white or bay. The earliest color illustrations of the golden variety show the males of one even golden bay without deviation, except in the tail, which seemed to be bronze and black. The neck of the female conformed in color to the neck of the male. The entire body was irregularly marked with black lines.

The present standard for color in the males is as follows: The hackle, back, saddle, shoulders, wing bows, breast, and under parts are silvery white; the wing coverts have the bottom web, or visible part, of each feather white, and the top web, or invisible part, of each feather coarsely penciled with dark; the secondaries are as white as possible, the top web being generally black or coarsely penciled; the tail is black tinged with green; the sickles and tail coverts are of a rich, transparent green surface color, black laced all around with a narrow stripe of white. In the female, the hackle is silvery white; the breast, thighs, back, saddle, shoulders, wing bows, wing coverts, tail, and tail coverts are silvery white, each feather being distinctly and evenly penciled straight across with fine parallel lines of a rich green hue—the penciling and the intervening ground color being of the same width; the secondaries should be penciled as much as possible, the markings a trifle coarse. The description embodied in both the American and English Standards conforms fairly well to the preceding description; but as seen in the exhibition pen, many of the males are striped about the fluff the same as in the golden penciled variety. Many of the males and females show markings in the hackle, and the black extends into the back of some of the males.

48. Mating.—The Silver Penciled Hamburg fowls must be mated the same as the Golden Penciled Hamburg. Although better offspring can be produced from them by single matings than is the case with the Golden Penciled Hamburg, the most beautiful fowls, especially females, are bred from exhibition hens of the highest quality mated to males that have been bred in line from such females mated to males that naturally come from them. To maintain the beautiful white plumage in males, exhibition males should be mated to females from a cockerel-breeding line.

SILVER SPANGLED HAMBURG

49. Origin.—The origin of the **Silver Spangled Hamburg** can best be discussed in connection with that of several of the other Hamburg varieties. For many years fowls known under the several names of Gold, Silver, and Black Pheasants, Gold and Silver Moonies, and Red Caps were bred in the northern part of England. All of these were undoubtedly of the same general character, differing in color only. When the poultrymen of England began to exhibit these fowls, the interest in them increased, and they were separated into the three

§7 BELGIAN, DUTCH, AND GERMAN FOWLS

varieties, the Silver Spangled, the Golden Spangled, and the Black Hamburg. The theory that the Black Hamburg came as black offspring from the Silver Moonies has been accepted. The Golden Spangled and the Silver Spangled Hamburgs were made by mingling the blood of Silver Pheasants with that of the Silver Moonies, thus producing better spangles in the silver variety. The Golden Pheasants and Golden Moonies were used for developing the golden variety. The Penciled Hamburg fowls were produced as distinctive offspring from the Bolton Grays and the Bolton Bays by selecting the best of them and breeding in line for more than fifty generations.

50. Development.—The Silver Spangled Hamburgs have been developed from the crude originals. More skill and more persistent attention have been given to the production of the several varieties of Hamburgs than has been expended on any other breed. As the result, a fowl has been created that has a well-established type and variety color, which is beyond comparison when of the best, but which reverts almost to the level of a mongrel when neglected.

51. History.—More attention was given in early days to the Hamburgs than was bestowed on any other breed of fowls, the Cochin excepted. In the publications of Wingfield & Johnson, in 1853, are shown a color illustration of Penciled Hamburg fowls and a black and white illustration of the spangled variety. Feathers illustrating their markings and a standard description of them were included in the publications. In the race for popularity, the Cochin was their rival from the beginning. The best fanciers have given great attention to the Hamburg fowls. English and Canadian fanciers have succeeded better with Hamburgs than have the fanciers of the United States, perhaps because they have given them more attention. Hamburg fowls have never been such favorites in America as in England. Boston, New York, and the larger exhibitions of Canada are the places in America where Hamburg fowls have been shown to any extent.

52. Description.—The Silver Spangled Hamburg fowls conform in every way to the description of the Golden Spangled

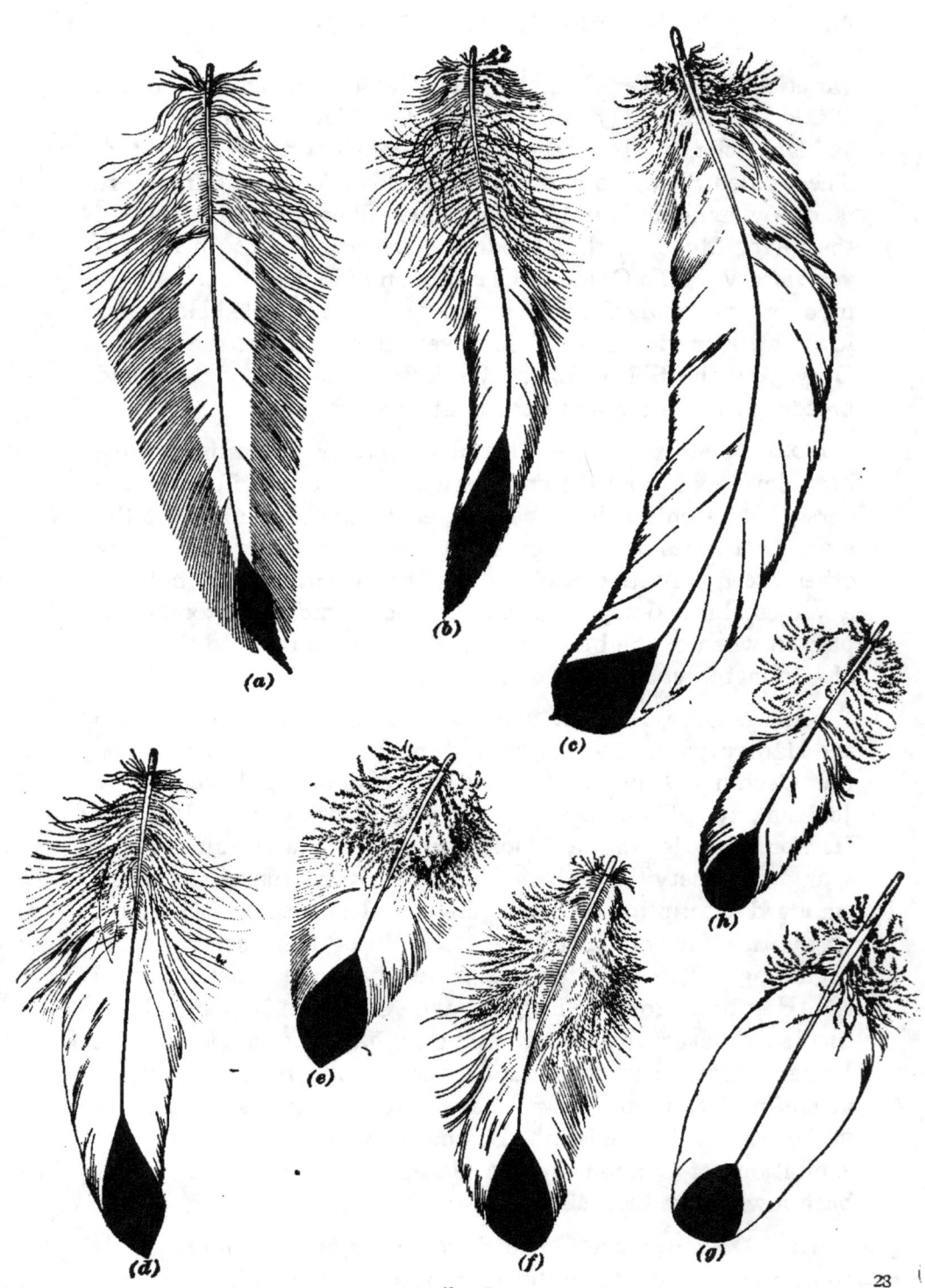

Fig. 5

§ 7 BELGIAN, DUTCH, AND GERMAN FOWLS

Hamburg, except that the body color of the Silver Spangled Hamburg is white and the spangles are black. The main difference will be found in the back and saddle. In the Silver Spangled Hamburg males, the feathers in these parts are white, with a small, black, dagger-shaped tip at the end. In the males of the Golden Spangled Hamburg, the feathers of the saddle are golden bay, each feather being striped down the center with green. The hackle of both males and females of the Silver Spangled Hamburg is silvery white, and each feather is ticked with a small, black, dagger-shaped tip. In the Golden Spangled Hamburg, the hackles of both males and females are marked down the center with a stripe of black.

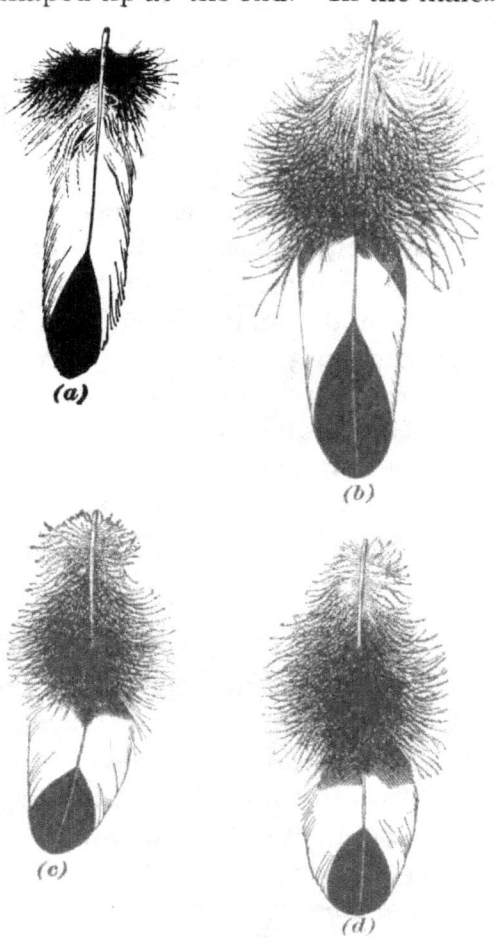

FIG. 6

53. Mating.—In the mating of Silver Spangled Hamburgs, fowls of the finest quality must be selected—those that have been bred in line from the best that can be produced. It is better to depend on well-selected fowls bred from the best breeding strains than to select from strains of unknown quality. A trio of Silver Spangled Hamburg fowls—composed of a male of the highest exhibition quality and two females, one of exhibition form and color and the other perhaps too dark for exhibition—selected from a strain known to be productive of the best, will answer well for producing both males and females of exhibition quality.

Sample feathers of Silver Spangled Hamburg males and females are shown in Figs. 5 and 6. In Fig. 5 (*a*) is shown a Silver Spangled Hamburg male hackle feather; in (*b*), a male saddle feather; in (*c*), a male main tail covert feather; in (*d*), a male lower tail covert; in (*e*), a male lower breast feather; in (*f*), a male thigh feather; in (*g*), a male secondary feather; and in (*h*), a male wing-bar feather. In Fig. 6 (*a*) is shown a Silver Spangled Hamburg female hackle feather; in (*b*), a female saddle feather; in (*c*), a female back feather; and in (*d*), a female breast feather.

No other color and markings are so difficult to produce as those of the plumage of the Silver Spangled Hamburg, and it is almost a necessity to establish a separate line of breeding for males and females, for only in this way can the best exhibition Hamburg fowls be produced.

These illustrations were made from feathers taken from Silver King and his mate, which were two noted prize winners. Although they had a much lighter shade of color in the under plumage than is demanded by the Standard, these two fowls were admitted to be among the best of their kind.

WHITE HAMBURG

54. Fowls having the same general formation as the silver penciled variety, and with white plumage, have been bred as a separate variety. They are supposed to have come as white offspring from the silver penciled variety, and are now known as the **White Hamburg.** Although they are recognized as a standard variety both in America and in England, they have never been bred to a perfection equal to that of the other varieties. This variety has the same kind of comb, face, wattles, and ear lobes as other Hamburg fowls; the beak, shanks, and toes are of a bluish shade, and the plumage is pure white to the skin, including the shafts of the feathers. To breed fowls of this variety of attractive quality requires careful selection for Hamburg type, Hamburg comb, and beautiful white ear lobes, clean, clear color in the shanks and toes, and the purest white plumage.

NON-STANDARD VARIETIES OF HAMBURG

55. There are two varieties of Hamburg besides those described, the **Buff Hamburg** and the **Speckled, or Mottled, Hamburg,** both of which may be considered as non-standard varieties. The Buff Hamburg was formerly recognized as a standard variety in England; the Speckled, or Mottled, Hamburg was not. The Buff Hamburg may have plumage of any color from lemon to rich golden buff. Of whatever shade the plumage color may be, it must be uniform throughout. It would be difficult to describe the difference between the Buff Hamburgs and the Rose-Comb Buff Leghorns were it not that the Hamburg has blue shanks and a more correct body formation. There are so few of the Buff Hamburgs as to make them almost obsolete. The Buff Hamburg fowls for breeding purposes must be carefully selected for Hamburg type and for an even shade of buff throughout. The color must be improved in the same way as for other buff fowls.

The Speckled, or Mottled, Hamburg has evidently been developed from badly marked offspring of the Spangled Hamburg. Fowls of this variety have been sparingly bred; they are never seen in America, and only occasionally in other countries. They have broken-colored plumage, black and white, more like the Mottled Java than the Houdan or Hamburg. They may be Mottled Javas with rose combs.

RED CAP

56. Origin.—It is claimed by early writers that the fowl from which the **Red Cap** fowls originated existed in the 14th century. Martin Doyle, who compiled an "Illustrated Book of Poultry," which is a revision of his book issued in 1854, gives as his authority for this statement Chaucer's description in "The Nonne's Preeste's Tale." In the early writings of Rev. E. S. Dixon, before quoted, Red Cap fowls are classed as a variety of Hamburg. The early writers that mention Red Cap fowls claim them to be mongrel Golden Spangled Hamburg.

57. Development.—Red Cap fowls received but little attention until after the Hamburg had been well developed, and the Red Cap fowls were taken up more as a separate breed than as a variety of the Hamburg. Following this, they were bred more carefully, and for a short time, beginning about 1890, or soon thereafter, they had a temporary popularity that attracted attention to them for a few years.

58. History.—Henry Belden, of England, one of the best informed poultrymen of his time, stated that Red Cap fowls were coarsely bred Golden Spangled Hamburg fowls; and that the size of the comb and body, rather than Hamburg proportion, had been developed. Other Englishmen prominent in poultry breeding agree with him; still others claim that they were produced by crossing Golden Spangled Hamburg with Old-English Game fowls. In early days, the Red Cap fowls were very plentiful in and about Yorkshire, England. They were bred almost exclusively for egg production and for market poultry. They were considered better for egg production at that time than the Leghorn fowls. They were known under the names of Pheasant fowls, Copper fowls, Yorkshire, and Derbyshire Red Caps. Very few of them have been kept in America.

59. Description.—The following description of Red Cap fowls was written for the "English Book of Poultry," 1902, by Albert E. Wragg, Edenson, Bakewell, England: "The Red Cap male is a fine-bodied bird of noble appearance, and nothing could be more ornamental than his symmetrically shaped comb full of a number of long spikes, with the leader behind. It should be well carried, firm and straight, and stand well off the eyes. The comb should be as large as can be comfortably carried by the bird. In size it should not greatly exceed $5\frac{1}{2}$ inches in length and $3\frac{1}{4}$ or 3 inches in width. The hen is shapely, very active, and a good forager; as a layer she is second to none."

The main character of the Red Cap is the immense rose comb, which stands high and has great width; the ear lobes are red. In the male, the hackle and saddle feathers are rich

§ 7 BELGIAN, DUTCH, AND GERMAN FOWLS

red, striped with black; the breast and tail are black; the back is red, marked with crescent-shaped black spangles. The hen's tail is black; the hackles are red, striped with black; the body plumage is reddish-brown, marked with black crescent-shaped spangles. The large-sized spangles resemble those found on the old Yorkshire Pheasants.

The American Standard describes the Red Caps as fowls of large size. The cock weighs $7\frac{1}{2}$ pounds; the cockerels and hens, 6 pounds; the pullets, 5 pounds. In form, they are Hamburg of large size. The males have horn-colored beaks; the eyes, face, comb, wattles, and ear lobes are red. According to the American Standard, the neck is blue-black, each feather being edged with red and the hackles shading off to black at the base. In the English Standard, the head is described as red, the hackles are red, each feather marked down the center with a stripe of black. According to both Standards, the back is red, spangled with black; the saddle is red, each feather being striped with black; the wing bows are a rich red; the coverts are red, each feather ending in a black spangle forming a black bar across the wing; the primaries and secondaries are red, tipped at the end with black; the breast and under body are black; and the tail and hangers are black, with a brilliant green sheen.

The Red Cap females are described as having the comb, face, wattles, ear lobes, and eyes, red; the beak, horn colored; the shanks and toes of both male and female, leaden blue or slate color. The head and hackle of the female is red, according to the English Standard, and brown, according to the American Standard; the back and breast, deep, rich reddish brown, free from smuttiness, each feather being tipped with a black or bluish-black crescent-shaped spangle. The color and markings of the breast, back, and wings should be as uniform as possible; the primaries and secondaries, according to the English Standard, are red, regularly tipped at the end with black; according to the American Standard, the primaries are dull black, with an edging of brown on the lower web; the secondaries have the upper web black, the lower web black, with an edging of brown, each feather being tipped with a black or a bluish-black spangle; the tail is black; the coverts are brown, ending with a black or bluish-black tip; the color of the under plumage, in both males and females, is dark or leaden blue. The combs are described as not so large as is mentioned by Mr. Wragg.

60. Mating.—The proper spangle of the Red Cap is crescentic in shape, not round. The spangles should be as dark as it is possible to have them; the body of the female should be a deep, rich, reddish brown, each feather not regularly spangled as the feathers of the Hamburg are, but irregularly tipped with crescent-shaped black tips of irregular size. The breast of the male is of a rich, glistening black with a purplish hue. In mating for the production of exhibition fowls, combs of large size should be encouraged, but, above all, they should be of perfect form and set straight on the head; no leaning to one side is permissible. Males of the best exhibition form and color mated to females of like character are the types that should be mated for the production of exhibition Red Cap fowls. Unless they are exquisite in form and have beautiful combs and glistening plumage, the colors of which stand out bold and true, they are of no value. To gain this, a strain must be as well established for them, such as will be needed for the production of Hamburg fowls of the finest type.

Red Cap fowls are worthy of more attention than they receive; they are excellent market poultry; they grow quickly into fairly good size, and are continual layers during winter and summer; the eggs are of good size and usually have white shells, though at times they are slightly tinted.

NON-STANDARD BREEDS OF DUTCH

61. The three breeds selected from among the several that are exclusive to their country are the *Breda*, the *Drente*, and the *Owl-Bearded Dutch*. The Breda might be classed as an Asiatic, the Drente as a Mediterranean, and the Owl-Bearded Dutch as a general-purpose fowl. All have white flesh and skin, dark or slate-blue shanks, and their eggs have white shells.

§ 7 BELGIAN, DUTCH, AND GERMAN FOWLS

All of them are bred in several varieties; the Drente is bred in all varieties common to the Mediterranean family.

Other Dutch fowls not mentioned resemble Polish and Penciled and Spangled Hamburgs. They are not known by these names, but in general appearance they are so much like inferior specimens of the fowls named that this seems to be nearest to a brief description that can be made of them.

BREDA

62. The **Breda** fowls originated in Holland. The fact that the males weigh from 6 to 9 pounds and the females from 5 to $6\frac{1}{2}$ pounds would indicate that they were largely descended from some one of the Asiatic family. This breed was at one time popular in Holland. A few of them were brought to America and are spoken of by Mr. Lewis, in his poultry book of 1871, as a fowl of medium size with a peculiar head, which was destitute of comb or crest, as shown in Fig. 7. They were bred in their native land for market purposes, but of late years they have been almost lost sight of. They are of several varieties, the most common of which are barred, black, blue, and pure white. They are some-

FIG. 7

what larger in size than are the Plymouth Rocks. They have been known in England and America as the Guelder or, as commonly called, Guelders. Early writers state that to produce barred or cuckoo color, the black Guelder cock should be mated with the white Guelder hen. They are sparingly feathered on the outside of the shanks and as pictured they were very full in breast, long in body, and resembled fowls of the Asiatic family to a slight degree.

DRENTE

63. The **Drente** fowls were originated in Holland, evidently from some of the Mediterranean fowls brought from Italy to that country. They have the general appearance of the Leghorn, and they have been bred in black, blue, cuckoo, gold and silver laced, gold and silver penciled, partridge, speckled, white, and yellow variety colors; yet none of them have been bred to a type or character that would make them attractive for exhibition purposes. They are prolific egg producers. In weight, the males average from 5 to $5\frac{1}{2}$ pounds; the females, from 4 to $4\frac{1}{2}$ pounds.

OWL-BEARDED DUTCH

64. The **Owl-Bearded Dutch** fowls might be likened to the Faverolle fowls. They were originated perhaps more than a century ago, although there does not seem to be any authentic record as to where or how they were made. The top of the head, the comb, and the beak of the Owl-Bearded Dutch are like the same parts of the La Flèche; the beard and the muff, like those parts of the Faverolle fowls. The male of the silver-laced variety has light top and dark under-body color somewhat like the Dorking; the female is marked like the laced Wyandotte. They have been known in four varieties: black, gold and silver laced, and white. They are somewhat smaller in size than are the Wyandotte, weighing at least a pound less. They are so little bred, even in their home country, as to make it quite a task to find any that are of fairly good quality.

GERMAN

65. Expanse of territory considered, Germany has fewer breeds or varieties of fowls than any other country. Most of the German breeds show the influence of the Italian varieties. Edward Brown mentions eight different kinds as belonging to Germany. Sketches made by Mr. Chatterton, of England, illustrate a number of kinds, all of which indicate a lack of careful breeding, and most of them show indications of having descended from some one of the Mediterranean varieties. In some of them may be noted the peculiar formation of head and comb found in the La Flèche; others plainly reveal the influence of Polish and Andulasian crosses. Of all the German fowls, the one breed that has found most consideration outside of its home country is the Lakenfelder.

LAKENFELDER

66. Origin.—The **Lakenfelder** fowl originated in Holland, and undoubtedly came from the same source as the Campines and Penciled Hamburgs. They may have come from the union of Campines of black-and-white plumage with white Leghorns, followed by selecting the best offspring and breeding them for form and color; but there is no absolute proof of such an origin.

67. Development.—The Lakenfelder fowls were formerly developed for egg production; later they became more attractive, and English and American fanciers have done much for their improvement.

68. History.—The Lakenfelder fowls were first mentioned as existing in West Hanover, where they were shown in 1835. They were bred in the same locality as Campine fowls, which were very dark in tail and hackle plumage. It has been mentioned by some that the dark Campines were bred with white

Italian fowls, and that in this way the Lakenfelder were made. They were first brought to England about 1900, and some were brought to America about the same time. The name Lakenfelder is referred to as meaning white spread over a black field (*laken*, lac or varnish; *feld*, field). Others refer to the name as meaning a shadow on a sheet, or black on white.

69. Description.—The Lakenfelder fowls have been recognized in England and a club standard description made for them; this standard represents them as having the neck of medium length, finely tapered and furnished with long, flowing hackle; skull, short; beak, strong; eyes, large, bright, and almond shaped; comb, single and of moderate size; wattles of medium length. The head and neck of both males and females are more like those of the Campines than those of the Leghorns. The body formation of both males and females is long and tapering to the tail. The breast is broad and full; the back is broad; the wings are of medium length; the tail is full. The sickles and coverts of the males are long and carried at an angle of about 45 degrees. The shanks and feet are of medium length, free from feathers; they have four toes. Their carriage is very sprightly. The males weigh from 5 to 6 pounds, and the females from $3\frac{1}{2}$ to $4\frac{1}{2}$ pounds. A peculiarity of this breed is the erect carriage of the comb of the females.

In color, the beak is dark; the eyes are red; comb, face, and wattles are bright red; the ear lobes are white; the shanks and feet are bluish or slate color; the plumage is black and white; the hackle and tail in both and the saddle hackle of the male are solid black, free from stripes or spots. The remainder of the plumage is pure white. The plumage is beautiful, its black being clean, clear, and glistening, and its white pure and unmixed. There should be a perfect separation of these two colors; neither should mar the beauty of the other.

70. Mating.—The only rule that can be followed for mating Lakenfelder fowls for the production of exhibition offspring is to select the best and have them conform as nearly as possible to the standard description of color. Great stress is laid on the presence of dark under plumage. This refers to the same color of under plumage necessary for producing proper surface color in Hamburg, Brahma, or Columbian Wyandotte fowls.

The difficult problem in breeding the Lakenfelder fowls is to keep the black and white separated and to have a black saddle without white on the male. This can be bred only after years of careful mating for a strain that will breed true to color and markings.

Silver-Spangled Hamburgs

Silver-Penciled Hamburgs

Golden Penciled Hamburgs

Silver Campines

BELGIAN, DUTCH, AND GERMAN FOWLS

Serial 1376   Edition 1

EXAMINATION QUESTIONS

Notice to Students.—*Study the Instruction Paper thoroughly before you attempt to answer these questions. Read each question carefully and be sure you understand it; then write the best answer you can. When your answers are completed, examine them closely, correct all the errors you can find, and see that every question is answered; then mail your work to us.*

(1) (*a*) Why are Braekel fowls better for general purposes than Campine fowls? (*b*) Describe the top color of the male Braekel.

(2) (*a*) From what districts of Belgium do the Campine fowls come? (*b*) What style of comb is correct for the Campine?

(3) (*a*) What body color is required for the male Campine under the English Club Standard? (*b*) What distinctive marking is on the tip of the feathers?

(4) (*a*) For what purpose are the Malines fowls bred? (*b*) Name the varieties bred in Belgium.

(5) Describe the peculiarity of the Turkey-Headed Malines.

(6) From what fowls have the modern Black Hamburg fowls been made?

(7) (*a*) From what locality were penciled fowls originally obtained? (*b*) Name two varieties of Penciled Hamburg fowls.

(8) (*a*) Describe the plumage color of the Golden Penciled Hamburg female. (*b*) Describe the plumage color of the Golden Penciled Hamburg male.

(9) (*a*) Describe the plumage color of Golden Spangled Hamburg male. (*b*) Describe the plumage color of Golden Spangled Hamburg female.

(10) Describe the plumage color of the Silver Penciled Hamburg male.

(11) Describe plumage color of the Silver Penciled Hamburg female.

(12) (*a*) From what sources were Silver Spangled Hamburg fowls obtained? (*b*) How does the color in the feathers shown in Figs. 5 and 6 compare with Standard requirements?

(13) How may a trio of Silver Spangled Hamburg fowls be made to produce both males and females for exhibition?

(14) How are White Hamburg fowls supposed to have come?

(15) (*a*) Where were Red Cap fowls plentifully bred in early days? (*b*) For what purpose were they bred?

(16) Describe the color and markings of the back and breast of the Red Cap.

(17) Describe the difference of color of the primaries and secondaries of Red Cap fowls in the English and American Standards.

(18) Where did Lakenfelder fowls originate?

(19) Describe the plumage color of the Lakenfelder.

(20) What are the difficult problems in breeding Lakenfelder fowls?

Mail your work on this lesson as soon as you have finished it and looked it over carefully. DO NOT HOLD IT until another lesson is ready.